Tucholsky Wagner Zola Scott Sydow Freud Schlegel
Turgenev Wallace Fonatne
Twain Walther von der Vogelweide Fouqué Friedrich II. von Preußen
Weber Freiligrath Frey
Fechner Fichte Weiße Rose von Fallersleben Kant Ernst Frommel
Richthofen
Engels Fielding Hölderlin Dumas
Fehrs Faber Flaubert Eichendorff Tacitus
Eliasberg Ebner Eschenbach
Feuerbach Maximilian I. von Habsburg Fock Zweig
Ewald Eliot Vergil
Goethe Elisabeth von Österreich London
Mendelssohn Balzac Shakespeare Dostojewski Ganghofer
Trackl Lichtenberg Rathenau Doyle Gjellerup
Mommsen Stevenson Hambruch
Thoma Tolstoi Lenz Droste-Hülshoff
Dach Verne von Arnim Hägele Hanrieder Humboldt
Reuter Hauff
Karrillon Rousseau Hagen Hauptmann Gautier
Garschin Defoe Baudelaire
Damaschke Hebbel
Descartes Hegel Kussmaul Herder
Wolfram von Eschenbach Dickens Schopenhauer Rilke George
Bronner Darwin Melville Grimm Jerome
Campe Horváth Aristoteles Bebel Proust
Bismarck Vigny Voltaire Federer Herodot
Barlach Heine
Gengenbach
Storm Casanova Tersteegen Grillparzer Georgy
Chamberlain Lessing Gilm
Langbein Gryphius
Brentano Lafontaine
Strachwitz Claudius Schiller Kralik Iffland Sokrates
Katharina II. von Rußland Bellamy Schilling
Gerstäcker Raabe Gibbon Tschechow
Löns Hesse Hoffmann Gogol Wilde Vulpius
Luther Heym Hofmannsthal Gleim
Roth Klee Hölty Morgenstern Goedicke
Luxemburg Heyse Klopstock Kleist
Puschkin Homer Mörike
La Roche Horaz Musil
Machiavelli Kierkegaard Kraft Kraus
Navarra Aurel Musset
Nestroy Marie de France Lamprecht Kind Kirchhoff Hugo Moltke
Laotse Ipsen Liebknecht
Nietzsche Nansen Ringelnatz
Marx Lassalle Gorki Klett Leibniz
von Ossietzky May
vom Stein Lawrence Irving
Petalozzi Knigge
Platon Pückler Michelangelo Kafka
Sachs Poe Kock
Liebermann
de Sade Praetorius Mistral Zetkin Korolenko

The publishing house tredition has created the series **TREDITION CLASSICS**. It contains classical literature works from over two thousand years. Most of these titles have been out of print and off the bookstore shelves for decades.

The book series is intended to preserve the cultural legacy and to promote the timeless works of classical literature. As a reader of a **TREDITION CLASSICS** book, the reader supports the mission to save many of the amazing works of world literature from oblivion.

The symbol of **TREDITION CLASSICS** is Johannes Gutenberg (1400 – 1468), the inventor of movable type printing.

With the series, tredition intends to make thousands of international literature classics available in printed format again – worldwide.

All books are available at book retailers worldwide in paperback and in hardcover. For more information please visit: www.tredition.com

tredition was established in 2006 by Sandra Latusseck and Soenke Schulz. Based in Hamburg, Germany, tredition offers publishing solutions to authors and publishing houses, combined with worldwide distribution of printed and digital book content. tredition is uniquely positioned to enable authors and publishing houses to create books on their own terms and without conventional manufacturing risks.

For more information please visit: www.tredition.com

The 'Pioneer': Light Passenger Locomotive of 1851 United States Bulletin 240, Contributions from the Museum of History and Technology, paper 42, 1964

John H. White

Imprint

This book is part of the TREDITION CLASSICS series.

Author: John H. White
Cover design: toepferschumann, Berlin (Germany)

Publisher: tredition GmbH, Hamburg (Germany)
ISBN: 978-3-8491-4776-1

www.tredition.com
www.tredition.de

Publications of the United States National Museum

The scholarly and scientific publications of the United States National Museum include two series, *Proceedings of the United States National Museum* and *United States National Museum Bulletin.*

In these series, the Museum publishes original articles and monographs dealing with the collections and work of its constituent museums—The Museum of Natural History and the Museum of History and Technology—setting forth newly acquired facts in the fields of anthropology, biology, history, geology, and technology. Copies of each publication are distributed to libraries, to cultural and scientific organizations, and to specialists and others interested in the different subjects.

The *Proceedings*, begun in 1878, are intended for the publication, in separate form, of shorter papers from the Museum of Natural History. These are gathered in volumes, octavo in size, with the publication date of each paper recorded in the table of contents of the volume.

In the *Bulletin* series, the first of which was issued in 1875, appear longer, separate publications consisting of monographs (occasionally in several parts) and volumes in which are collected works on related subjects. *Bulletins* are either octavo or quarto in size, depending on the needs of the presentation. Since 1902 papers relating to the botanical collections of the Museum of Natural History have been published in the *Bulletin* series under the heading *Contributions from the United States National Herbarium*, and since 1959, in *Bulletins* titled "Contributions from the Museum of History and Technology," have been gathered shorter papers relating to the collections and research of that Museum.

The present collection of Contributions, Papers 34-44, comprises Bulletin 240. Each of these papers has been previously published in separate form. The year of publication is shown on the last page of each paper.

Frank A. Taylor
Director, United States National Museum

[Pg 241]

6

Contributions from
The Museum of History and Technology:
Paper 42

The "Pioneer": Light Passenger Locomotive of 1851 In the Museum of History and Technology

John H. White

Contents

Figure 1.—The "Pioneer," built in 1851, shown here as renovated and exhibited in the Museum of History and Technology, 1964. In 1960 the locomotive was given to the Smithsonian Institution by the Pennsylvania Railroad through John S. Fair, Jr. (Smithsonian photo 63344B.)

[Pg 243] *John H. White*

The "PIONEER":
LIGHT PASSENGER LOCOMOTIVE
of 1851
In the Museum of History and Technology

In the mid-nineteenth century there was a renewed interest in the light, single-axle locomotives which were proving so very successful for passenger traffic. These engines were built in limited number by nearly every well-known maker, and among the few remaining is the 6-wheel "Pioneer,"

11

on display in the Museum of History and Technology, Smithsonian Institution. This locomotive is a true representation of a light passenger locomotive of 1851 and a historic relic of the mid-nineteenth century.

The Author: *John H. White is associate curator of transportation in the Smithsonian Institution's Museum of History and Technology.*

The "Pioneer" is an unusual locomotive and on first inspection would seem to be imperfect for service on an American railroad of the 1850's. This locomotive has only one pair of driving wheels and no truck, an arrangement which marks it as very different from the highly successful standard 8-wheel engine of this period. All six wheels of the *Pioneer* are rigidly attached to the frame. It is only half the size of an 8-wheel engine of 1851 and about the same size of the 4−2−0 so common in this country some 20 years earlier. Its general arrangement is that of the rigid English locomotive which had, years earlier, proven unsuitable for use on U.S. railroads.

These objections are more apparent than real, for the *Pioneer*, and other engines of the same design, proved eminently successful when used in the service for which they were built, that of light passenger traffic. The *Pioneer's* rigid wheelbase is no problem, for when it is compared to that of an 8-wheel engine it is found to be about four feet less; and its small size is no problem when we realize it was not intended for heavy service. Figure 2, a diagram, is a comparison of the *Pioneer* and a standard 8-wheel locomotive.

Since the service life of the *Pioneer* was spent on the Cumberland Valley Railroad, a brief account of that line is necessary to an understanding of the service history of this locomotive.

[Pg 244]

Exhibits of the "Pioneer"

The *Pioneer* has been a historic relic since 1901. In the fall of that year minor repairs were made to the locomotive so that it might be used in the sesquicentennial celebration at Carlisle, Pennsylvania. On October 22, 1901, the engine was ready for service, but as it neared Carlisle a copper flue burst. The fire was extinguished and the *Pioneer* was pushed into town by another engine. In the twentieth century, the *Pioneer* was displayed at the Louisiana Purchase Exposition, St. Louis, Missouri, in 1904, and at the Wheeling, West

Virginia, semicentennial in 1913. In 1927 it joined many other historic locomotives at the Baltimore and Ohio Railroad's "Fair of the Iron Horse" which commemorated the first one hundred years of that company. From about 1913 to 1925 the *Pioneer* also appeared a number of times at the Apple-blossom Festival at Winchester, Virginia. In 1933-1934 it was displayed at the World's Fair in Chicago, and in 1948 at the Railroad Fair in the same city. Between 1934 and March 1947 it was exhibited at the Franklin Institute, Philadelphia, Pennsylvania.

The Cumberland Valley Railroad

The Cumberland Valley Railroad (C.V.R.R.) was chartered on April 2, 1831, to connect the Susquehanna and Potomac Rivers by a railroad through the Cumberland Valley in south-central Pennsylvania. The Cumberland Valley, with its rich farmland and iron-ore deposits, was a natural north-south route long used as a portage between these two rivers. Construction began in 1836, and because of the level valley some 52 miles of line was completed between Harrisburg and Chambersburg by November 16, 1837. In 1860, by way of the Franklin Railroad, the line extended to Hagerstown, Maryland. It was not until 1871 that the Cumberland Valley Railroad reached its projected southern terminus, the Potomac River, by extending to Powells Bend, Maryland. Winchester, Virginia, was entered in 1890 giving the Cumberland Valley Railroad about 165 miles of line. The railroad which had become associated with the Pennsylvania Railroad in 1859, was merged with that company in 1919.

By 1849 the Cumberland Valley Railroad was in poor condition; the strap-rail track was worn out and new locomotives were needed. Captain Daniel Tyler was hired to supervise rebuilding the line with T-rail, and easy grades and curves. Tyler recommended that a young friend of his, Alba F. Smith, be put in charge of modernizing and acquiring new equipment. Smith recommended to the railroad's Board of Managers on June 25, 1851, that "much lighter engines than those now in use may be substituted for the passenger transportation and thereby effect a great saving both in point of fuel and road repairs...." [1] Smith may well have gone on to explain that the road was operating 3- and 4-car passenger trains with a locomotive weighing about 20 tons; the total weight was about 75 tons, equalling the uneconomical deadweight of 1200 pounds per passenger. Since speed was not an important consideration (30 mph being a good average), the use of lighter engines would improve the deadweight-to-passenger ratio and would not result in a slower schedule.

The Board of Managers agreed with Smith's recommendations and instructed him "... to examine the two locomotives lately built by Mr. Wilmarth and now in the [protection?] of Captain Tyler at Norwich and if in his judgment they are adequate to our wants ... have them forwarded to the road." [2] Smith inspected the locomotives not long after this resolution was passed, for they were on the road by the time he made the following report [3] to the Board on September 24, 1851:

In accordance with a resolution passed at the last meeting of your body relative to the small engines built by Mr. Wilmarth I proceeded to Norwich to make trial of their capacity—fitness or suitability to the Passenger transportation of our Road—and after as thorough a trial as circumstances would admit (being on another Road than our own) I became satisfied that with some necessary improvements which would not be expensive (and are now being made at our shop) [Pg 245] the engines would do the business of our Road not only in a manner satisfactory in point of speed and certainty but with greater ultimate economy in Expenses than has before been practised in this Country.

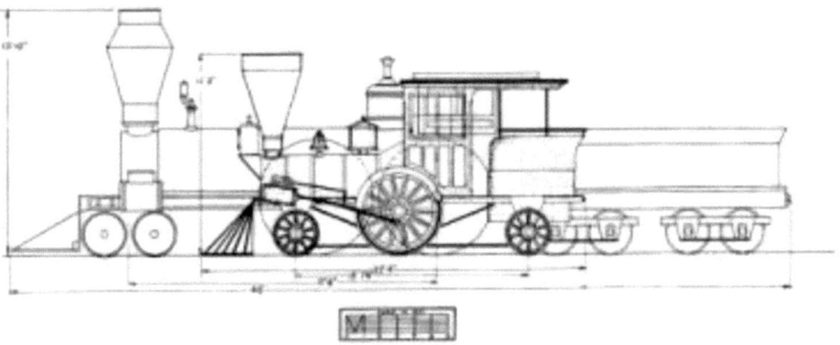

Figure 2.—Diagram comparing the *Pioneer* (shaded drawing) with the *Columbia*, a standard 8-wheel engine of 1851. (Drawing by J. H. White.)

Columbia

Hudson River Railroad
Lowell Machine Shop, 1852
Wt. 27-1/2 tons (engine
only)

Pioneer

Cumberland Valley Railroad
Seth Wilmarth, 1851
12-1/2 tons

Cyl. 16-1/2 x 22 inches   8-1/2 x 14 inches
Wheel diam. 84 inches   54 inches

After making the above trial of the Engines—I stated to your Hon. President the result of the trial—with my opinion of their Capacity to carry our passenger trains at the speed required which was decidedly in favor of the ability of the Engines. He accordingly agreed that the Engines should at once be forwarded to the Road in compliance with the Resolution of your Board. I immediately ordered the Engines shipped at the most favorable rates. They came to our Road safely in the Condition in which they were shipped. One of the Engines has been placed on the Road and I believe performed in such a manner as to convince all who are able to judge of this ability to perform—although the maximum duty of the Engines was not performed on account of some original defects which are now being remedied as I before stated.

Within ten days the Engine will be able to run regularly with a train on the Road where in shall be enabled to judge correctly of their merits.

An accident occurred during the trial of the Small Engine at Norwich which caused a damage of about $300 in which condition the Engine came here and is now being repaired—the cost of which will be presented to your Board hereafter. As to the fault or blame of parties connected with the accident as also the question of responsibility for Repairs are questions for your disposal. I therefore leave the matter until further called upon.

The Expenses necessarily incurred by the trial of the Engines and also the Expenses of transporting the same are not included in the Statement herewith presented, the whole amount of which will not probably exceed $400.00.

These two locomotives became the Cumberland Valley Railroad's *Pioneer* (number 13) and *Jenny Lind* (number 14). While Smith notes that one of the engines was damaged during the inspection trials, Joseph Winters, an employee of the Cumberland Valley who claimed he was accompanying the engine enroute to Chambersburg at the time of their delivery, later recalled that both engines were damaged in transit. [4] According to Winters a train ran into the rear of the *Jenny Lind*, damaging both it and the *Pioneer*, the accident occurring near Middletown, Pennsylvania. The *Jenny Lind* was repaired at Harrisburg but the *Pioneer*, less seriously damaged, was taken for repairs to the main shops of the Cumberland Valley road at Chambersburg.

[Pg 246]

Figure 3.—"Pioneer," about 1901, showing the sandbox and large headlamp. Note the lamp on the cab roof, now used as the headlight. (Smithsonian photo 49272.)

While there seems little question that these locomotives were not built as a direct order for the Cumberland Valley Railroad, an article [5] appearing in the *Railroad Advocate* in 1855 credits their design to Smith. The article speaks of a 2—2—4 built for the Macon and Western Railroad and says in part:

This engine is designed and built very generally upon the ideas, embodied in some small tank engines designed by A. F. Smith, Esq., for the Cumberland Valley road. Mr. Smith is a strong advocate of light engines, and his novel style and proportions of engines, as built for him a few years since, by Seth Wilmarth, at Boston, are

18

known to some of our readers. Without knowing all the circumstances under which these engines are worked on the Cumberland
Valley road, we should not venture to repeat all that we have heard
of their performances, it is enough to say that they are said to do
more, in proportion to their weight, than any other engines now in
use.

The author believes that the *Railroad Advocate's* claim of Smith's
design of the *Pioneer* has been confused with his design of the *Utility*
(figs. 6, 7). Smith designed this compensating-lever engine to haul
trains over the C.V.R.R. bridge at Harrisburg. It was built by Wilmarth in 1854.

Figure 4. — Map of the Cumberland Valley Railroad as it appeared in 1919.

Figure 5.—An early broadside of the Cumberland Valley Railroad.

According to statements of Smith and the Board of Managers quoted on page 244, the *Pioneer* and the *Jenny Lind* were not new when purchased from their maker, Seth Wilmarth. Although of recent manufacture, previous to June 1851, they were apparently doing service on a road in Norwich, Connecticut. It [Pg 247] should be mentioned that both Smith and Tyler were formerly associated with the Norwich and Worcester Railroad and they probably learned of these two engines through this former association. It is possible that the engines were purchased from Wilmarth by the Cumberland Valley road, which had bought several other locomotives from Wilmarth in previous years. It was the practice of at least one other New England engine builder, the Taunton Locomotive Works, to manufacture engines on the speculation that a buyer would be found; if no immediate buyers appeared the engine was leased to a local road until a sale was made. [6]

Regarding the *Jenny Lind* and *Pioneer*, Smith reported [7] to the Board of Managers at their meeting of March 17, 1852:

The small tank engines which were purchased last year ... and which I spoke in a former report as undergoing at that time some necessary improvements have since that time been fairly tested as to their capacity to run our passenger trains and proved to be equal to the duty.

The improvements proposed to be made have been completed only on one engine [*Jenny Lind*] which is now running regularly with passenger trains—the cost of repairs and improvements on this engine (this being the one accidentally broken on the trial) amounted to $476.51. The other engine is now in the shop, not yet ready for service but will be at an early day.

[Pg 248]

22

Figure 6.—The "Utility" as rebuilt to an 8-wheel engine, about 1863 or 1864. It was purchased by the Carlisle Manufacturing Co. in 1882 and was last used in 1896. (Smithsonian photo 36716F.)

Figure 7.—The "Utility," designed by Smith A. F. and constructed by Seth Wilmarth in 1854, was built to haul trains across the bridge at Harrisburg, Pa.

[Pg 249]

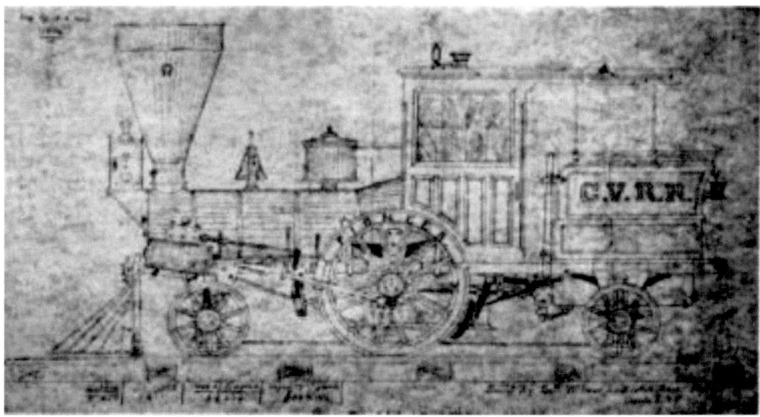

Figure 8.—The earliest known illustration of the *Pioneer*, drawn by A. S. Hull, master mechanic of the Cumberland Valley Railroad

in 1876. It depicts the engine as it appeared in 1871. (*Courtesy of Paul Westhaeffer.*)

The *Pioneer* and *Jenny Lind* achieved such success in action that the president of the road, Frederick Watts, commented on their performance in the annual report of the Cumberland Valley Railroad for 1851. Watts stated that since their passenger trains were rarely more than a baggage car and two coaches, the light locomotives "... have been found to be admirably adapted to our business." The Cumberland Valley Railroad, therefore, added two more locomotives of similar design in the next few years. These engines were the *Boston* and the *Enterprise,* also built by Wilmarth in 1854-1855.

Watts reported the *Pioneer* and *Jenny Lind* cost $7,642. A standard 8-wheel engine cost about $6,500 to $8,000 each during this period. In recent years, the Pennsylvania Railroad has stated the *Pioneer* cost $6,200 in gold, but is unable to give the source for this information. The author can discount this statement for it does not seem reasonable that a light, cheap engine of the pattern of the *Pioneer* could cost as much as a machine nearly twice its size.

Figure 9.—Annual pass of the Cumberland Valley Railroad issued in 1863.

[Pg 250]

24

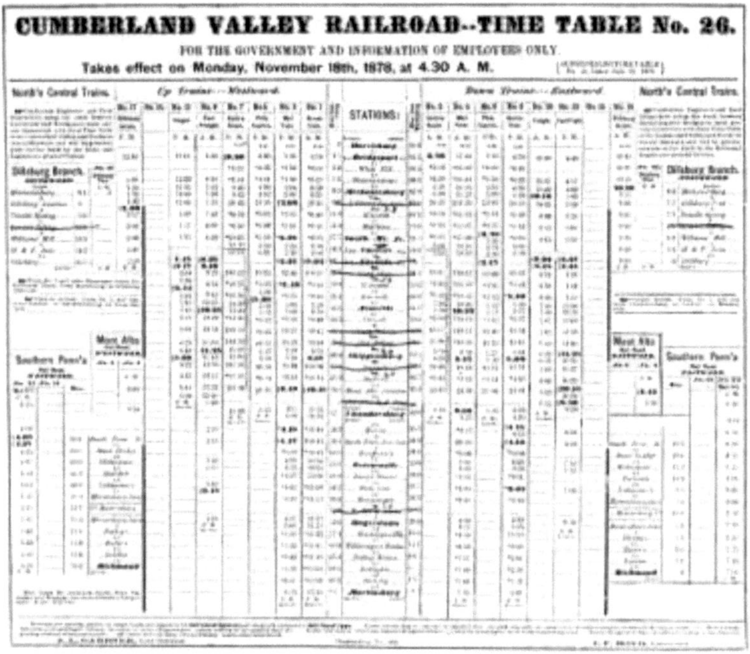

CUMBERLAND VALLEY RAILROAD--TIME TABLE No. 26.

FOR THE GOVERNMENT AND INFORMATION OF EMPLOYEES ONLY.

Takes effect on Monday, November 18th, 1878, at 4.30 A. M.

Figure 10.—Timetable of the Cumberland Valley Railroad for 1878.

Service History of the *Pioneer*

After being put in service, the *Pioneer* continued to perform well and was credited as able to move a 4-car passenger train along smartly at 40 mph. [8] This tranquility was shattered in October 1862 by a raiding party led by Confederate General J. E. B. Stuart which [Pg 251] burned the Chambersburg shops of the Cumberland Valley Railroad. The *Pioneer, Jenny Lind,* and *Utility* were partially destroyed. The Cumberland Valley Railroad in its report for 1862 stated:

The Wood-shop, Machine-shop, Black-smith-shop, Engine-house, Wood-sheds, and Passenger Depot were totally consumed, and with the Engine-house three second-class Engines were much injured by the fire, but not so destroyed but that they may be restored to usefulness.

However, no record can be found of the extent or exact nature of the damage. The shops and a number of cars were burned so it is reasonable to assume that the cab and other wooden parts of the locomotive were damaged. One unverified report in the files of the Pennsylvania Railroad states that part of the roof and brick wall fell on the *Pioneer* during the fire causing considerable damage. In June 1864 the Chambersburg shops were again burned by the Confederates, but on this occasion the railroad managed to remove all its locomotives before the raid. During the Civil War, the Cumberland Valley Railroad was obliged to operate longer passenger trains to satisfy the enlarged traffic. The *Pioneer* and its sister single-axle engines were found too light for these trains and were used only on work and special trains. Reference to table 1 will show that the mileage of the *Pioneer* fell off sharply for the years 1860-1865.

Table 1.—Yearly Mileage of the Pioneer
(From Annual Reports of the Cumberland Valley Railroad)

Year:	*Miles*
1852	[a] 3,182
1853	[b] 20,722
1854	18,087

Year	Mileage
1855	14,151
1856	20,998
1857	22,779
1858	29,094
1859	29,571
1860	4,824
1861	4,346
1862	([c])
1863	5,339
1864	224
1865	2,215
1866	20,546
1867	5,709
1868	13,626
1869	1,372
1870	...
1871	2,102
1872	4,002
1873	3,721
1874	3,466
1875	636
1876	870
1877	406
1878	4,433
1879	...
1880	8,306
1881	([d])
Total	[e] 244,727

[a] Mileage 1852 for January to September (no record of mileage recorded in Annual Reports previous to 1852).

[b] 15,000 to 20,000 miles per year was considered very high mileage for a locomotive of the 1850's.

[c] No mileage reported for any engines due to fire.

[d] Not listed on roster.

[e] The Pennsylvania Railroad claims a total mileage of 255,675. This may be accounted for by records of mileages for 1862, 1870, and 1879.

In 1871 the *Pioneer* was remodeled by A. S. Hull, master mechanic of the railroad. The exact nature of the alterations cannot be determined, as no drawings or photographs of the engine previous to this time are known to exist. In fact, the drawing (fig. 8) prepared by Hull in 1876 to show the engine as remodeled in 1871 is the oldest known illustration of the *Pioneer*. Paul Westhaeffer, a lifelong student of Cumberland Valley R. R. history, states that according to an interview with one of Hull's descendants the only alteration made to the *Pioneer* during the 1871 "remodeling" was the addition of a handbrake. The road's annual report of 1853 describes the *Pioneer* as a six-wheel tank engine. The report of 1854 mentions that the *Pioneer* used link motion. These statements are enough to give substance to the idea that the basic arrangement has survived unaltered and that it has not been extensively rebuilt, as was the *Jenny Lind* in 1878.

By the 1870's, the *Pioneer* was too light for the heavier cars then in use and by 1880 it had reached the end of its usefulness for regular service. After nearly thirty years on the road it had run 255,675 miles. Two new passenger locomotives were purchased in 1880 to handle the heavier trains. In 1881 the *Pioneer* was dropped from the roster, but was used until about 1890 for work trains. After this time it was stored in a shed at Falling Spring, Pennsylvania, near the Chambersburg yards of the C.V.R.R.

Mechanical Description of the *Pioneer*

Figure 11.—"Pioneer," about 1901, scene unknown. (*Photo courtesy of Thomas Norrell.*)

After the early 1840's the single-axle locomotive, having one pair of driving wheels, was largely [Pg 252] superseded by the 8-wheel engine. The desire to operate longer trains and the need for engines of greater traction to overcome the steep grades of American roads called for coupled driving wheels and machines of greater weight than the $4-2-0$. After the introduction of the $4-4-0$, the single-axle engine received little attention in this country except for light service or such special tasks as inspection or dummy engines.

Figure 12.—The "Pioneer" in Carlisle, Pa., 1901. (*Photo courtesy of Thomas Norrell.*)

There was, however, a renewed interest in "singles" in the early 1850's because of W. B. Adams' experiments with light passenger locomotives in England. In 1850 Adams built a light single-axle tank locomotive for the Eastern Counties Railway which proved very economical for light passenger traffic. It was such a success that considerable interest in light locomotives was generated in this country as well as in England. Nearly 100 single-axle locomotives were built in the United States between about 1845-1870. These engines were built by nearly every well-known maker, from Hinkley in Boston to the Vulcan Foundry in San Francisco. Danforth Cooke & Co. of Paterson built a standard pattern 4−2−4 used by many roads. One of these, the *C. P. Huntington*, survives to the present time.

The following paragraphs describe the mechanical details of the *Pioneer* as it appears on exhibition in the Smithsonian Institution's new Museum of History and Technology.

BOILER

The boiler is the most important and costly part of a steam locomotive, representing one-fourth to one-third of the total cost. A poorly built or designed boiler will produce a poor locomotive no matter how well made the remainder of mechanism. The boiler of the *Pioneer* is of the wagon-top, crownbar, fire-tube [Pg 253] style and is made of a 5/16-inch thick, wrought-iron plate. The barrel is very small, in keeping with the size of the engine, being only 27 inches in diameter. While some readers may believe this to be an extremely early example of a wagon-top boiler, we should remember that most New England builders produced few locomotives with the Bury (dome) boiler and that the chief advocates of this later style were the Philadelphia builders. By the early 1850's the Bury boiler passed out of favor entirely and the wagon top became the standard type of boiler with all builders in this country.

Sixty-three iron tubes, 1-7/8 inches by 85 inches long are used. The original tubes may have been copper or brass since these were easier to keep tight than the less malleable iron tubes. The present tube sheet is of iron but was originally copper. Its thickness cannot be conveniently measured, but it is greater than that of the boiler shell, probably about 1/2 to 5/8 inch. While copper tubes and tube sheets were not much used in this country after about 1870, copper was employed as recently as 1950 by Robert Stephenson & Hawthorns, Ltd., on some small industrial locomotives.

The boiler shell is lagged with wooden tongue-and-groove strips about 2-1/2 inches wide (felt also was used for insulation during this period). The wooden lagging is covered with Russia sheet iron which is held in place and the joints covered by polished brass bands. Russia sheet iron is a planish iron having a lustrous, metallic gray finish.

Alba F. Smith

Alba F. Smith, the man responsible for the purchase of the *Pioneer*, was born in Lebanon, Connecticut, June 28, 1817. [9] Smith showed promise as a mechanic at an early age and by the time he was 22 had established leadpipe works in Norwich. His attention was drawn particularly to locomotives since the tracks of the Norwich and Worcester Railroad passed his shop. His attempts to develop a spark arrester for locomotives brought Smith to the favorable attention of Captain Daniel Tyler (1799-1882), president of the Norwich and Worcester Railroad. When Tyler was hired by the Cumberland Valley Railroad in 1850 to supervise the line's rebuilding, he persuaded the managers of that road to hire Smith as superintendent of machinery. [10] Smith was appointed as superintendent of the machine shop of the Cumberland Valley Railroad on July 22, 1850. [11] On January 1, 1851, he became superintendent of the road.

In March of 1856 Smith resigned his position with the Cumberland Valley Railroad and became superintendent of the Hudson River Railroad, where he remained for only a year. During that time he designed the coal-burning locomotive *Irvington*, rebuilt the Waterman condensing dummy locomotive for use in hauling trains through city streets, and developed a superheater. [12]

After retiring from the Hudson River Railroad he returned to Norwich and became active in enterprises in that area, including the presidency of the Norwich and Worcester Railroad. While the last years of Smith's life were devoted to administrative work, he found time for mechanical invention as well. In 1862 he patented a safety truck for locomotives, and became president of a concern which controlled the most important patents for such devices. [13] Alba F. Smith died on July 21, 1879, in Norwich, Connecticut.

[Pg 254]

Figure 13.—Advertisement of Seth Wilmarth appearing in Boston city directory for 1848-1849.

[Pg 255]

36

Figure 14.—The "Fury," built for the Boston and Worcester Railroad in 1849 by Wilmarth. It was known as a "Shanghai" because of its great height. (Smithsonian Chaney photo 6443.)

Figure 15.—The "Neptune," built for the Boston and Worcester in 1847 by Hinkley and Drury. Note the similarity of this engine and the *Fury*.

Figure 16.—The "Pioneer" as first exhibited in the Arts and Industries building of the Smithsonian Institution prior to restoration of the sandbox. (Smithsonian photo 48069D.)

[BOILER continued]

The steam dome (fig. 18) is located directly over the firebox, inside the cab. It is lagged and jacketed in an identical manner to the boiler. The shell of the dome is of 5/16-inch wrought iron, the top cap is a cast-iron plate which also serves as a manhole cover offering access to the boiler's interior for inspection and repair.

[Pg 256]

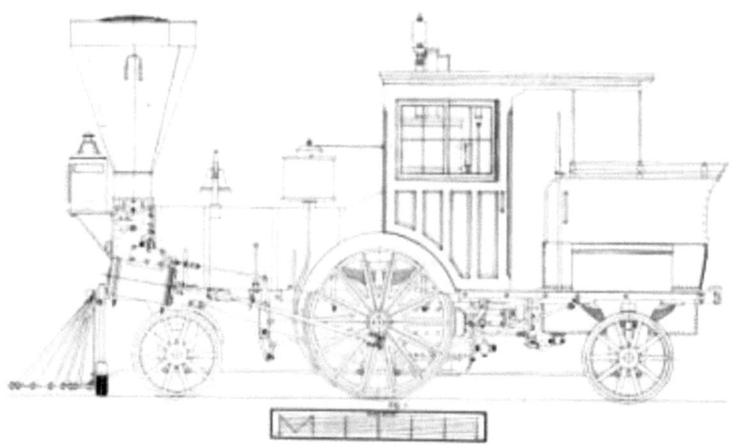

Figure 17.—"Pioneer" locomotive. (Drawing by J. H. White.)

[Pg 257]

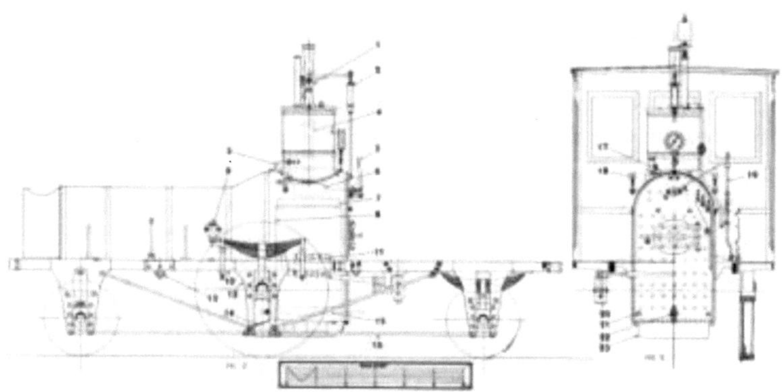

Figure 18.—"Pioneer" locomotive, (1) Safety valve, (2) spring balance, (3) steam jet, (4) dry pipe, (5) throttle lever, (6) throttle, (7) crown bar, (8) front tube sheet, (9) check valve, (10) top rail, (11) rear-boiler bracket, (12) pedestal, (13) rocker bearing, (14) damper, (15) grate, (16) bottom rail, (17) pump heater valve, (18) cylinder lubricator, (19) reversing lever, (20) brake shoe, (21) mud ring, (22) blowoff cock, (23) ashpan. (Drawing by J. H. White.)

A round plate, 20 inches in diameter, riveted on the forward end of the boiler, just behind the bell stand, was found when the old jacket was removed in May 1963. The size and shape of the hole, which the plate [Pg 258] covers, indicate that a steam dome or manhole was located at this point. It is possible that this was the original location of the steam dome since many builders in the early 1850's preferred to mount the dome forward of the firebox. This was done in the belief that there was less danger of priming because the water was less agitated forward of the firebox.

The firebox is as narrow as the boiler shell and fits easily between the frame. It is a deep and narrow box, measuring 27 inches by 28 inches by about 40 inches deep, and is well suited to burning wood. A deep firebox was necessary because a wide, shallow box suitable for coal burning, allowed the fuel to burn so quickly it was difficult to fire the engine effectively. With the deep, narrow firebox, wood was filled up to the level of the fire door. In this way, the fire did not burn so furiously and did not keep ahead of the fireman; at the same time, since it burned so freely, a good fire was always on hand. The *Pioneer* burned oak and hickory. [14] For the firebox 5/16-inch thick sheet was used, for heavier sheet would have blistered and flaked off because of the intense heat of the fire and the fibrous quality of wrought-iron sheet of the period. Sheet iron was fabricated from many small strips of iron rolled together while hot. These strips were ideally welded into a homogeneous sheet, but in practice it was found the thicker the sheet the less sure the weld.

The fire grates are cast iron and set just a few inches above the bottom of the water space so that the water below the grates remains less turbulent and mud or other impurities in the water settle here. Four bronze mud plugs and a blowoff cock are fitted to the base of the firebox so that the sediment thus collected can be removed (figs. 17, 18).

The front of the boiler is attached to the frame by the smokebox, which is a cylinder, bolted on a light, cast-iron saddle (not part of the cylinder castings nor attached to them, but bolted directly to the top rail of the frame; it may be a hastily made repair put on at the shops of the C.V.R.R.). The rear of the boiler is attached to the frame by two large cast-iron brackets, one on each side of the firebox (fig.

18). These are bolted to the top rail of the frame but the holes in the brackets are undoubtedly slotted, so that they may slide since the boiler will expand about 1/4 inch when heated. In addition to the crown bars, which strengthen the crown sheet, the boiler is further strengthened by stay bolts and braces located in the wagon top over the firebox, where the boiler had been weakened by the large hole necessary for the steam dome. This boiler is a remarkably light, strong, and compact structure.

BOILER FITTINGS

Few boiler fittings are found on the *Pioneer* and it appears that little was done to update the engine with more modern devices during its many years of service. With the exception of the steam gauge, it has no more boiler fitting than when it left the builder's shop in 1851.

Figure 19.—Backhead of the *Pioneer*. (Smithsonian photo 48069F.)

The throttle valve is a simple slide valve and must have been primitive for the time, for the balance-poppet throttle valve was in use in this country previous to 1851. It is located directly below the steam dome even though it was common practice to place the throttle valve at the front of the boiler in the smokebox. Considering the cramped condition inside the smokebox, there would seem to be little space for the addition of the throttle valve; hence its present location. The dry pipe projects up into the steam dome to gather the hottest, driest steam for the cylinders. The inverted, funnel-like cap on the top of the dry pipe is to prevent priming, as drops of water may travel up the sides of the pipe and then to the cylinders, with the possibility of great damage. After the steam enters the throttle valve it passes through the front end of the valve, through the top of the boiler via the dry pipe (fig. 18), through the front tube sheet, and then to the cylinders via the petticoat pipes. The throttle lever is a simple arrangement readily understood from the drawings. It has no latch and the throttle lever is held in any desired setting by the wingnut and quadrant shown in figure 18. The water level in the boiler is indicated by the three brass cocks located on the backhead. No gauge glass is used; they were not employed in this country until the 1870's, although they were commonly used in England at the time the *Pioneer* was built.

While two safety valves were commonly required, only one was used on the *Pioneer*. The safety valve is located on top of the steam dome. Pressure is exerted on the lever by a spring balance, fixed at the forward end by a knife-blade bearing. The pressure can be adjusted by the thumbscrew on the balance. The graduated scale on the balance gave a general but uncertain indication of the boiler pressure. The valve itself is a poppet held against the face of the valve seat by a second knife blade attached to the [Pg 259] lever. The ornamental column forming the stand of the safety valve is cast iron and does much to decorate the interior of the cab. The pipe carrying the escaping steam projects through the cab roof. It is made of copper with a decorative brass band. This entire mechanism was replaced by a modern safety valve for use at the Chicago Railroad Fair (1949). Fortunately, the old valve was preserved and has since been replaced on the engine.

The steam gauge is a later addition, but could have been put on as early as the 1860's, since the most recent patent date that it bears is 1859. It is an Ashcroft gauge having a handsome 4—4—0 locomotive engraved on its silver face.

The steam jet (item 3, fig. 18) is one of the simplest yet most notable boiler fitting of the *Pioneer*, being nothing more than a valve tapped into the base of the steam dome with a line running under the boiler jacket to the smokestack. When the valve is opened a jet of steam goes up the stack, creating a draft useful for starting the fire or enlivening it as necessary. This device was the invention of Alba F. Smith in 1852, according to the eminent 19th-century technical writer and engineer Zerah Colburn. [15]

The two feedwater pumps (fig. 20) are located beneath the cab deck (1, fig. 17). They are cast-iron construction and are driven by an eccentric on the driving-wheel axle (fig. 27). The airchamber or dome (1, fig. 27) imparts a more steady flow of the water to the boiler by equalizing the surges of water from the reciprocating pump plunger. A steam line (3, fig. 18), which heats the pump and prevents freezing in cold weather, is regulated by a valve in the cab (figs. 18, 27). Note that the line on the right side of the cab has been disconnected and plugged.

The eccentric drive for the pumps is unusual, and the author knows of no other American locomotive so equipped. Eastwick and Harrison, it is true, favored an eccentric drive for feed pumps, but they mounted the eccentric on the crankpin of the rear driving wheel and thus produced in effect a half-stroke pump. This was not an unusual arrangement, though a small crank was usually employed in place of the eccentric. The full-stroke crosshead pump with which the *Jenny Lind* (fig. 22) is equipped, was of course the most common style of feed pump used in this country in the 19th century.

Of all the mechanisms on a 19th-century locomotive, the feed pump was the most troublesome. If an engineer could think of nothing else to complain about, he could usually call attention to a defective pump and not be found a liar. Because of this, injectors were adopted after their introduction in 1860. It is surprising that the *Pioneer*, which was in regular service as late as 1880 and has been

under <u>steam</u> many times since for numerous exhibitions, was never
fitted with one of these devices. Because its stroke is [Pg 260] short
and the plunger is in less rapid motion, the present eccentric arrangement is more complex but less prone to disorder than the sim-
pler but faster crosshead pump.

Seth Wilmarth

Little is known of the builder of the *Pioneer*, Seth Wilmarth, and nothing in the way of a satisfactory history of his business is available. For the reader's general interest the following information is noted. [16]

Seth Wilmarth was born in Brattleboro, Vermont, on September 8, 1810. He is thought to have learned the machinist trade in Pawtucket, Rhode Island, before coming to Boston and working for the Boston Locomotive Works, Hinkley and Drury proprietors. In about 1836 he opened a machine shop and, encouraged by an expanding business, in 1841 he built a new shop in South Boston which became known as the Union Works. [17] Wilmarth was in the general machine business but his reputation was made in the manufacture of machine tools, notably lathes. He is believed to have built his first locomotive in 1842, but locomotive building never became his main line of work. Wilmarth patterned his engines after those of Hinkley and undoubtedly, in common with the other New England builders of this period, favored the steady-riding, inside-connection engines. The "Shanghais," so-called because of their great height, built for the Boston and Worcester Railroad by Wilmarth in 1849, were among the best known inside-connection engines operated in this country (fig. 14). While the greater part of Wilmarth's engines was built for New England roads, many were constructed for lines outside that area, including the Pennsylvania Railroad, Ohio and Pennsylvania Railroad, and the Erie.

A comparison of the surviving illustrations of Hinkley and Wilmarth engines of the 1850's reveals a remarkable similarity in their details (figs. 14 and 15). Notice particularly the straight boiler, riveted frame, closely set truck wheels, feedwater pump driven by a pin on the crank of the driving wheel, and details of the dome cover. All of the features are duplicated exactly by both builders. This is not surprising considering the proximity of the plants and the fact that Wilmarth had been previously employed by Hinkley.

In 1854 Wilmarth was engaged by the New York and Erie Railroad to build fifty 6-foot gauge engines. [18] After work had been started on these engines, and a large store of material had been purchased for their construction, Wilmarth was informed that the railroad could not pay cash but that he would have to take notes in payment. [19] There was at this time a mild economic panic and notes could be sold only at a heavy discount. This crisis closed the Union Works. The next year, 1855, Seth Wilmarth was appointed master mechanic of the Charlestown Navy Yard, Boston, where he worked for twenty years. He died in Malden, Massachusetts, on November 5, 1886.

[Pg 261]

[BOILER FITTINGS continued]

Figure 20.—Feedwater pump of the *Pioneer*. (Smithsonian photo 63344.)

The check valves are placed slightly below the centerline of the boiler (fig. 18). These valves are an unfinished bronze casting and appear to be of a recent pattern, probably dating from the 1901 renovation. At the time the engine was built, it was usual to house these valves in an ornamental spun-brass casing. The smokestack is of the bonnet type commonly used on wood-burning locomotives in this country between about 1845 and 1870. The exhaust steam from the cylinders is directed up the straight stack (shown in phantom in fig. 27) by the blast pipe. This creates a partial vacuum in the smokebox that draws the fire, gases, ash, and smoke through the boiler tubes from the firebox. The force of the exhausting steam blows them out the stack. At the top of the straight stack is a deflecting cone which slows the velocity of the exhaust and changes its direction causing it to go down into the funnel-shaped outer casing of the stack. Here, the heavy embers and cinders are collected and prevented from directly discharging into the countryside as dangerous firebrands. Wire netting is stretched overtop of the deflecting cone to catch the lighter, more volatile embers which may defy the action of the cone. The term "bonnet stack" results from the fact that this netting is similar in shape to a lady's bonnet. The cinders thus accumulated in the stack's hopper could be emptied by opening a plug at the base of the stack.

While the deflecting cone was regarded highly as a spark arrester and used practically to the exclusion of any other arrangement, it had the basic defect of keeping the smoke low and close to the train. This was a great nuisance to passengers, as the low trailing smoke blew into the cars. If the exhaust had been allowed to blast straight out the stack high into the air, most of the sparks would have burned out before touching the ground.

[Pg 262]

Figure 21.—"Pioneer" on exhibit in old Arts and Industries building of the Smithsonian Institution. In this view can be seen the bonnet screen of the stack and arrangement of the boiler-frame braces and other details not visible from the floor. (Smithsonian photo 48069A.)

Figure 22.—"Jenny Lind," sister engine of the *Pioneer*, shown here as rebuilt in 1878 for use as an inspection engine. It was scrapped in March 1905. (*Photo courtesy of E. P. Alexander.*)

[Pg 263]

Figure 23.—Cylinder head with valve box removed.

Figure 24.—Bottom of valve box with slide valve removed.

Figures 25 and 26.—Cylinder with valve box removed, showing valve face.

FRAME

The frame of the *Pioneer* defies an exact classification but it more closely resembles the riveted- or sandwich-type frame than any other (figs. 18, 27). While the simple bar frame enjoyed the greatest popularity in the last century, riveted frames were widely used in this country, particularly by the New England builders between about 1840 and 1860. The riveted frame was fabricated from two plates of iron, about 5/8-inch thick, cut to the shape of the top rail and the pedestal. A bar about 2 inches square was riveted between the two plates. A careful study of photographs of Hinkley and other New England-built engines of the period will reveal this style of construction. The frame of the *Pioneer* differs from the usual riveted frame in that the top rail is 1-3/4 inches thick by 4-1/8 inches deep and runs the length of the locomotive. The pedestals are made of two 3/8-inch plates flush-riveted to each side of the top rail. The cast-iron shoes which serve as guides for the journal boxes also act as spacers between the pedestal plates.

The bottom rail of the frame is a 1-1/8-inch diameter rod which is forged square at the pedestals and forms the pedestal cap. The frame is further stiffened by two diagonal rods running from the top of each truck-wheel pedestal to the base of the driving-wheel pedestal, forming a truss. Six rods, riveted to the boiler shell and bolted to the frame's top rail, strengthen the frame laterally. Four of these rods can be seen easily as they run from the frame to the middle of the boiler; the other two are riveted to the underside of the boiler. The attachment of these rods to the boiler was an undesirable practice, for the boiler shell [Pg 264] was thus subjected to the additional strain of the locomotive's vibrations as it passed over the road. In later years, as locomotives grew in size, this practice was avoided and frames were made sufficiently strong to hold the engine's machinery in line without using the boiler shell.

The front and rear frame beams are of flat iron plate bolted to the frame. The rear beam had been pushed in during an accident, and instead of its being replaced, another plate was riveted on and bent out in the opposite direction to form a pocket for the rear coupling pin. Note that there is no drawbar and that the coupler is merely bolted to the beams. Since the engine only pulled light trains, the arrangement was sufficiently strong.

RUNNING GEAR

The running gear is simply sprung with individual leaf springs for each axle; it is not connected by equalizing levers. To find an American locomotive not equipped with equalizers is surprising since they were almost a necessity to produce a reasonably smooth ride on the rough tracks of American railroads. Equalizers steadied the motion of the engine by distributing the shock received by any one wheel or axle to all the other wheels and axles so connected, thus minimizing the effects of an uneven roadbed. The author believes that the *Pioneer* is a hard-riding engine.

The springs of the main drives are mounted in the usual fashion. The rear boiler bracket (fig. 18) is slotted so that the spring hanger may pass through for its connection with the frame. The spring of the leading wheels is set at right angles to the frame (fig. 27) and bears on a beam, fabricated of iron plate, which in turn bears on the journal boxes. The springs of the trailing wheels are set parallel with the frame and are mounted between the pedestal plates (fig. 18).

The center of the driving wheel is cast iron and has spokes of the old rib pattern, which is a T in cross section, and was used previous to the adoption of the hollow spoke wheel. In the mid-1830's Baldwin and others used this rib-pattern style of wheel, except that the rib faced inside. The present driving-wheel centers are unquestionably original. The sister engine *Jenny Lind* (fig. 22) was equipped with identical driving wheels. The present tires are very thin and beyond their last turning. They are wrought iron and shrunk to fit the wheel centers. Flush rivets are used for further security. The left wheel, shown in figure 17, is cracked at the hub and is fitted with an iron ring to prevent its breaking.

The truck wheels, of the hollow spoke pattern, are cast iron with chilled treads. They were made by Asa Whitney, one of the leading car-wheel manufacturers in this country, whose extensive plant was located in Philadelphia. Made under Whitney's patent of 1866, these wheels may well have been added to the *Pioneer* during the 1871 rebuilding. Railroad wheels were not cast from ordinary cast iron, which was too weak and brittle to stand the severe service for which they were intended, but from a high-quality cast iron similar to that used for cannons. Its tensile strength, which ranged from 31,000 to 36,000 psi, was remarkably high and very nearly approached that of the best wrought-iron plate.

The cylinders are cast iron with an 8-1/2-inch bore about half the size of the cylinders of a standard 8-wheel engine. The cylinders are bolted to the frame but not to the saddle, and are set at a 9° angle to clear the leading wheels and at the same time to line up with the center of the driving-wheel axle. The wood lagging is covered with a decorative brass jacket. Ornamental brass jacketing was extensively used on mid-19th-century American locomotives to cover not only the cylinders but steam and sand boxes, check valves, and valve boxes. The greater expense for brass (Russia iron or painted sheet iron were a cheaper substitute) was justified by the argument that brass lasted the life of the engine, and could be reclaimed for scrap at a price approaching the original cost; and also that when brightly polished it reflected the heat, preventing loss by radiation, and its bright surface could be seen a great distance, thus helping to prevent accidents at grade crossings. The reader should be careful not to misconstrue the above arguments simply as rationalization on the part of master mechanics more intent on highly decorative machines than on the practical considerations involved.

The valve box, a separate casting, is fastened to the cylinder casting by six bolts. The side cover plates when removed show only a small opening suitable for inspection and adjustment of the valve. The valve box must be removed to permit repair or removal of the valve. A better understanding of this mechanism and the layout of the parts can be gained from a study of figures 23-26, 28 (8, 8A, and 8B).

[Pg 265]

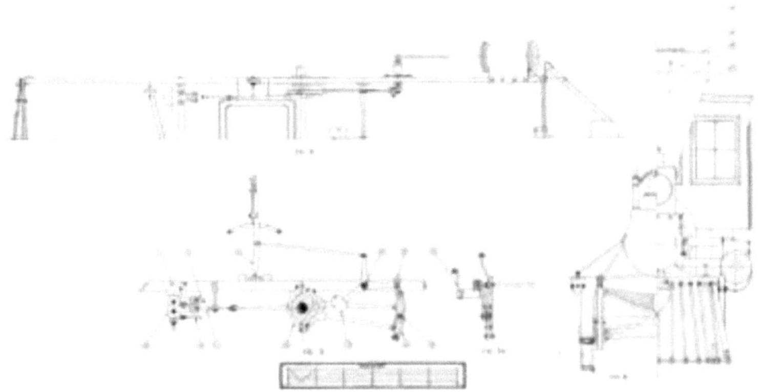

Figure 27.—"Pioneer" locomotive. (1) Air chamber, (2) reversing lever, (3) counterweight, (4) reversing shaft, (5) link hanger, (6)

rocker, (7) feedwater line to boiler, (8) link block, (9) link, (10) eccentric, (11) pump plunger, (12) pump steamheater line, (13) feedwater pump, (14) wire netting [bonnet], (15) deflecting cone, (16) stack, (17) stack hopper. (Drawing by J. H. White.)

[Pg 266]

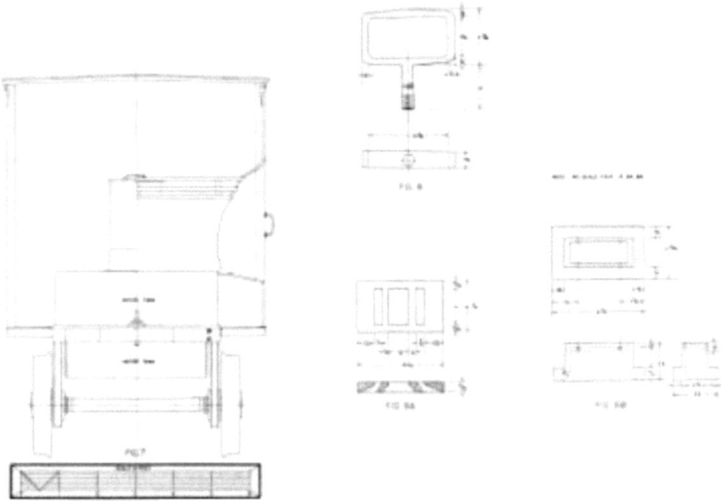

Figure 28.—Rear elevation of *Pioneer* and detail of valve shifter; valve face and valve. (Drawing by J. H. White.)

Both crossheads were originally of cast iron but one of these has been replaced and is of steel. They run into steel guides, bolted at the forward end to the rear cylinder head and supported in the rear by a yoke. The yoke is one of the more finished and better made pieces on the entire engine (fig. 27). The main rod is of the old pattern, round in cross section, and only 1-1/2 inches in diameter at the largest point.

VALVE GEAR

The valve gear is of the Stephenson shifting-link pattern (see fig. 27), a simple and dependable motion used extensively in this country between about 1850 and 1900. The author believes that this is the original valve gear of the *Pioneer*, since the first mention (1854) in

56

the *Annual Report* of the Cumberland Valley Railroad of the style of valve gear used by each engine, states that the *Pioneer* was equipped with a shifting-link motion. Assuming this to be the original valve gear of the *Pioneer*, it must be regarded as an early application, because the Stephenson motion was just being introduced into American locomotive practice in the early 1850's. Four eccentrics drive the motion; two are for forward motion and two for reverse. The link is split and made of two curved pieces. The rocker is fabricated of several forged pieces keyed and bolted together. On better made engines the rocker would be a one-piece forging. The lower arm of each rocker is curiously shaped, made with a slot so that the link block may be adjusted. Generally, the only adjustment possible was effected by varying the length of the valve stem by the adjusting nuts provided. A simple weight and lever attached to the reversing shaft serve as a counterbalance for the links and thus assist the engineer in shifting the valve motion. There are eight positions on the quadrant of the reversing lever.

Figure 29.—"Pioneer" on exhibit in old Arts and Industries building, showing the tank and backhead. (Smithsonian photo 48069E.)

MISCELLANEOUS NOTES

The cab is solid walnut with a natural finish. It is very possible that the second cab was added to the locomotive after the 1862 fire. A brass gong used by the [Pg 267] conductor to signal the engineer is fastened to the underside of the cab roof. This style of gong was in use in the 1850's and may well be original equipment.

The water tank is in two sections, one part extending below the deck, between the frame. The tank holds 600 gallons of water. The tender holds one cord of wood.

The small pedestal-mounted sandbox was used on several Cumberland Valley engines including the *Pioneer*. This box was removed from the engine sometime between 1901 and 1904. It was on the engine at the time of the Carlisle sesquicentennial but disappeared by the time of the St. Louis exposition. Two small sandboxes, mounted on the driving-wheel splash guards, replaced the original box. The large headlamp (fig. 3) apparently disappeared at the same time and was replaced by a crudely made lamp formerly mounted on the cab roof as a backup light. Headlamps of commercial manufacture were carefully finished and made with parabolic reflectors, elaborate burners, and handsomely fitted cases. Such a lamp could throw a beam of light for 1000 feet. The present lamp has a flat cone-shaped piece of tin for a reflector.

The brushes attached to the pilot were used in the winter to brush snow and loose ice off the rail and thus improve traction. In good weather the brushes were set up to clear the tracks.

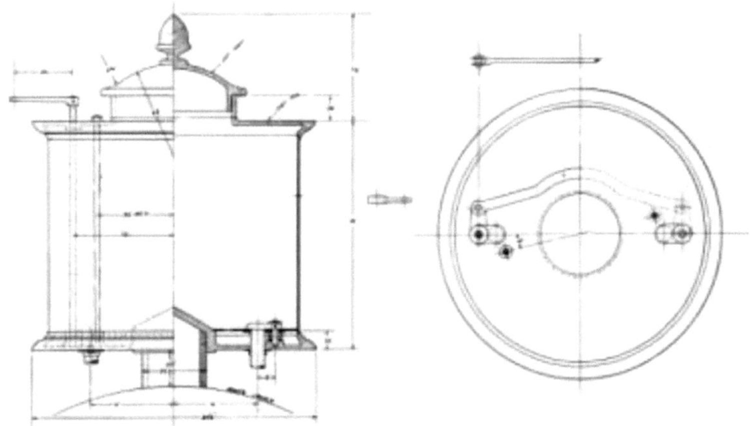

Figure 30.—Reconstructed sandbox replaced on the locomotive, August 1962. (Drawing by J. H. White.)

After the *Pioneer* had come to the National Museum, it was decided that some refinishing was required to return it as nearly as possible to the state of the original engine. Replacing the sandbox was an obvious change. [20] The brass cylinder jackets were also replaced. The cab was stripped and carefully refinished as natural wood. The old safety valve was replaced, [Pg 268] as already mentioned. Rejacketing the boiler with simulated Russia iron produced a most pleasing effect, adding not only to the authenticity of the display but making the engine appear lighter and relieving the somber blackness which was not characteristic of a locomotive of the 1850's. Several minor replacements are yet to be done; chiefly among these are the cylinder-cock linkage and a proper headlamp.

The question arises, has the engine survived as a true and accurate representation of the original machine built in 1851? In answer, it can be said that although the *Pioneer* was damaged en route to the Cumberland Valley Railroad, modified on receipt, burned in 1862, and operated for altogether nearly 40 years, surprisingly few new appliances have been added, nor has the general arrangement been changed. Undoubtedly, the main reason the engine is so little changed is that its small size and odd framing did not invite any large investment for extensive alteration for other uses. But there can be no positive answer as to its present variance from the original appearance as represented in the oldest known illustration of it—the Hull drawing of 1871 (fig. 8). There are few, if any, surviving 19th-century locomotives that have not suffered numerous rebuildings and are not greatly altered from the original. The *John Bull*, also in the U.S. National Museum collection, is a good example of a machine many times rebuilt in its 30 years of service. [21] Unless other information is uncovered to the contrary, it can be stated that the *Pioneer* is a true representation of a light passenger locomotive of 1851.

Footnotes

[1] *Minutes of the Board of Managers of the Cumberland Valley Railroad.* This book may be found in the office of the Secretary, Pennsylvania Railroad, Philadelphia, Pa., June 25, 1851. Hereafter cited as "Minutes C.V.R.R."

[2] Ibid.

[3] Minutes C.V.R.R.

[4] *Franklin Repository* (Chambersburg, Pa.), August 26, 1909.

[5] *Railroad Advocate* (December 29, 1855), vol. 2, p. 3.

[6] C. E. Fisher, "Locomotives of the New Haven Railroad, " *Railway and Locomotive Historical Society Bulletin* (April 1938), no. 46, p. 48.

[7] Minutes C.V.R.R.

[8] *Evening Sentinel* (Carlisle, Pa.), October 23, 1901.

[9] *Norwich Bulletin* (Norwich, Conn.), July 24, 1879. All data regarding A. F. Smith is from this source unless otherwise noted.

[10] *Railway Age* (September 13, 1889), vol. 14, no. 37. Page 600 notes that Tyler worked on C.V.R.R. 1851-1852; Smith's obituary (footnote 9) mentions 1849 as the year; and minutes of C.V.R.R. mention Tyler as early as 1850.

[11] Minutes C.V.R.R.

[12] A. F. Holley, *American and European Railway Practice* (New York: 1861). An illustration of Smith's superheater is shown on plate 58, figure 13.

[13] John H. White, "Introduction of the Locomotive Safety Truck," (Paper 24, 1961, in *Contributions from the Museum of History and Technology: Papers 19-30*, U.S. National Museum Bulletin 228; Washington: Smithsonian Institution, 1963), p. 117.

[14] *Annual Report*, C.V.R.R., 1853.

[15] Zerah Colburn, *Recent Practice in Locomotive Engines* (1860), p. 71.

[16] *Railroad Gazette* (September 27, 1907), vol. 43, no. 13, pp. 357-360. These notes on Wilmarth locomotives by C. H. Caruthers were printed with several errors concerning the locomotives of the Cumberland Valley Railroad and prompted the preparation of these present remarks on the history of Wilmarth's activities. Note that on page 359 it is reported that only one compensating-lever engine was built for the C.V.R.R. in 1854, and not two such engines in 1852. The *Pioneer* is incorrectly identified as a "Shanghai," and as being one of three such engines built in 1871 by Wilmarth.

[17] The author is indebted to Thomas Norrell for these and many of the other facts relating to Wilmarth's Union Works.

[18] *Railroad Gazette* (October 1907), vol. 43, p. 382.

[19] *Boston Daily Evening Telegraph* (Boston, Mass.), August 11, 1854. The article stated that one engine a week was built and that 10 engines were already completed for the Erie. Construction had started on 30 others.

[20] The restoration work has been ably handled by John Stine of the Museum staff. Restoration started in October 1961.

[21] S. H. Oliver, *The First Quarter Century of the Steam Locomotive in America* (U.S. National Museum Bulletin 210; Washington: Smithsonian Institution, 1956), pp. 38-46.